Speed Arithmetic: A Formula Handbook

N.B. Singh

DEDICATION

To Nature,

I dedicate this book to you, the source of all life. You are my inspiration, my teacher, and my friend.

Thank you for teaching me about the beauty of the world around me. Thank you for showing me the power of the natural world. Thank you for giving me a sense of peace and tranquillity.

I promise to do my part to protect you and your many wonders. I will teach my children about the importance of conservation and sustainability. I will work to make the world a better place for all living things.

Thank you for everything, Nature.

With love,

N.B Singh

Contents

Preface

Welcome to "Speed Arithmetic: A Formula Handbook." This handbook is designed to be your comprehensive guide to mastering the art of mental calculation. In a world dominated by technology, the ability to perform rapid and accurate arithmetic mentally remains a valuable skill. Whether you're a student, professional, or simply an enthusiast, this handbook provides a rich collection of formulas, techniques, and strategies to enhance your mental arithmetic abilities.

Why Speed Arithmetic Matters

In today's fast-paced environment, the demand for quick and precise calculations is ever-present. Speed arithmetic not only saves time but also cultivates mental agility, strategic thinking, and problem-solving skills. Whether you're tackling complex equations or making swift calculations in daily life, the formulas presented in this handbook are designed to make your mental arithmetic journey seamless and efficient.

What You Will Find in This Handbook

The handbook is structured to take you on a progressive journey through various aspects of speed arithmetic. It begins with a solid foundation in basic arithmetic operations, gradually introducing advanced techniques, specialized systems like Vedic Mathematics and the Trachtenberg System, memory techniques, practical applications, and speed arithmetic competitions.

Key Features

- Comprehensive coverage of basic and advanced arithmetic operations.

- Exploration of specialized systems for unique perspectives on mental calculation.

- Mnemonics and memory techniques to aid in recalling formulas and concepts.

- Practical applications to demonstrate the relevance of speed arithmetic in everyday scenarios.

- Speed arithmetic competitions to test and enhance your skills under time constraints.

How to Use This Handbook

This handbook is designed for both learning and reference. Feel free to read it cover to cover for a comprehensive understanding, or use it as a quick reference guide for specific formulas and techniques. Each chapter builds on the previous ones, offering a systematic approach to mastering speed arithmetic.

Enjoy Your Journey

I hope this handbook becomes a valuable companion on your journey to mastering speed arithmetic. May your calculations be swift, accurate, and enjoyable!

Chapter 1

Introduction

1.1 Background

Speed arithmetic has been a subject of fascination and practical importance throughout history. The need for rapid calculations arises in various fields, from everyday tasks to professional endeavors. Mathematicians and scholars have developed numerous techniques to enhance computational speed, efficiency, and accuracy. In this section, we explore the historical context and significance of speed arithmetic.

1.1.1 Historical Development

Speed arithmetic techniques have roots in ancient civilizations, where scholars devised clever methods for calculations. One notable example is the ancient Indian system of Vedic Mathematics, which introduced sutras (formulae) for rapid mental calculations. Over time, these techniques evolved, incorporating insights from various cultures and mathematical traditions.

1.1.2 Importance in Modern Times

In today's fast-paced world, the ability to perform calculations swiftly is invaluable. Professionals such as accountants, engineers, and data analysts benefit from mastering speed arithmetic. Moreover, students can improve their mathematical skills and boost exam performance by adopting these techniques.

1.1.3 Formula: Two-Digit Multiplication

One fundamental speed arithmetic formula is the technique for multiplying two-digit numbers. Let $AB \times CD$ be the multiplication of two-digit numbers, where A, B, C, and D are digits. The result can be calculated using the following formula:

$$(10A + B) \times (10C + D) = 100AC + 10(AD + BC) + BD$$

Example:

Let's multiply 23 by 34 using the formula:

$$(10 \times 2 + 3) \times (10 \times 3 + 4) = 100 \times 2 \times 3 + 10 \times (2 \times 4 + 3 \times 3) + 3 \times 4 = 782$$

In this way, complex multiplication can be simplified using the formula, demonstrating the efficiency of speed arithmetic techniques.

Real-world Application:

Consider a scenario where a shopkeeper needs to calculate the total cost of 47 items priced at 32 each. Using the two-digit multiplication formula, the shopkeeper can quickly determine the total cost without relying on a calculator.

$$(10 \times 4 + 7) \times (10 \times 3 + 2) = 100 \times 4 \times 3 + 10 \times (4 \times 2 + 7 \times 3) + 7 \times 2 = 1504$$

In this real-world application, speed arithmetic proves its practical utility in everyday situations.

Chapter 2

Basic Arithmetic

2.1 Addition

The process of addition is fundamental to arithmetic, and speed arithmetic offers efficient techniques to perform addition swiftly. In this section, we explore various methods and formulas for rapid addition.

2.1.1 Column Addition

Traditional column addition involves aligning numbers by place value and adding digits column by column, carrying over when necessary. A key speed arithmetic formula simplifies this process:

$$
\begin{array}{r}
ABC \\
+\,DEF \\
\hline
XYZ
\end{array}
$$

The result X, Y, and Z can be calculated as follows:

$$Z = (C + F) \bmod 10$$

$$Y = (B + E + (C + F) \bmod 10) \bmod 10$$

$$X = (A + D + (B + E + (C + F) \operatorname{div} 10) \operatorname{div} 10)$$

Example:

Consider the addition of 456 and 789:

$$
\begin{array}{r}
456 \\
+\ 789 \\
\hline
1245
\end{array}
$$

The calculations would be:

$$Z = (6 + 9) \bmod 10 = 5$$

$$Y = (5 + 8 + (6 + 9) \operatorname{div} 10) \bmod 10 = 4$$

$$X = (4 + 7 + (5 + 8 + (6 + 9) \operatorname{div} 10) \operatorname{div} 10) = 1$$

Therefore, $456 + 789 = 1245$.

2.1.2   Cross Addition

Cross addition is a speed arithmetic technique where corresponding digits of the numbers are added diagonally. This method eliminates the need for carrying over, simplifying the mental calculation process.

Example:

Let's cross add 456 and 789:

$$
\begin{array}{r}
4 \\
00 \\
05 \\
+\,000 \\
04 \\
+\,000 \\
01 \\
+\,000 \\
456 \\
+\,789 \\
\hline
1245
\end{array}
$$

In cross addition, we add the numbers along the diagonals, and the result is obtained without carrying over.

2.2 Subtraction

Subtraction, a fundamental arithmetic operation, is an essential skill in various fields. Speed arithmetic techniques provide efficient methods for performing subtraction quickly and accurately. In this section, we explore different strategies and formulas for rapid subtraction.

2.2.1 Column Subtraction

Traditional column subtraction involves aligning numbers by place value and subtracting digits column by column, borrowing when necessary. A key speed arithmetic formula simplifies this process:

$$ABC$$
$$- DEF$$
$$\overline{}$$
$$XYZ$$

The result X, Y, and Z can be calculated as follows:

$$Z = (C - F) \bmod 10$$

$$Y = (B - E - (C - F) \operatorname{div} 10) \bmod 10$$

$$X = (A - D - (B - E - (C - F) \operatorname{div} 10) \operatorname{div} 10)$$

Example:

Consider the subtraction of 789 from 456:

$$456$$
$$- 789$$
$$\overline{}$$
$$XYZ$$

The calculations would be:

$$Z = (6 - 9) \bmod 10 = 7$$

$$Y = (5 - 8 - (6 - 9) \operatorname{div} 10) \bmod 10 = 7$$

$$X = (4 - 7 - (5 - 8 - (6 - 9) \operatorname{div} 10) \operatorname{div} 10) = 6$$

Therefore, $456 - 789 = 667$.

2.2.2 Complementary Subtraction

Complementary subtraction is a speed arithmetic technique that transforms a subtraction problem into an addition problem by using complements. It simplifies mental calculations and reduces the need for borrowing.

Example:

Let's subtract 789 from 456 using complementary subtraction:

$$\begin{array}{r} 456 \\ -\,789 \\ \hline XYZ \end{array}$$

To use complements, we find the complement of the number to be subtracted, which is 789. The complement is 211 (since $789 + 211 = 1000$). Now, we add 456 and 211:

$$\begin{array}{r} 456 \\ +\,211 \\ \hline XYZ \end{array}$$

The result is $XYZ = 667$, which is the same as the previous example, confirming the accuracy of complementary subtraction.

2.3 Multiplication

Multiplication is a fundamental operation in arithmetic, and speed arithmetic techniques provide efficient methods for performing multiplications rapidly. In this section, we explore different strategies and formulas for quick multiplication.

2.3.1 Two-Digit Multiplication

A common and essential speed arithmetic technique is two-digit multiplication. For two-digit numbers AB and CD, where A, B, C, and D are digits, the result can be calculated using the following formula:

$$\begin{bmatrix} & A & B \\ \times & C & D \\ \hline & AC & BD \\ & AD & BC \\ \hline & XYZ & \end{bmatrix}$$

Where X, Y, and Z are obtained by summing the products of the crosswise digits.

Example:

Let's multiply 23 by 34 using the formula:

$$\begin{bmatrix} & 2 & 3 \\ \times & 3 & 4 \\ \hline & 6 & 8 \\ & 1 & 2 \\ \hline & 7 & 8 \end{bmatrix}$$

Therefore, $23 \times 34 = 782$.

2.3.2   Vedic Mathematics: Nikhilam Sutra

The Nikhilam Sutra is a Vedic Mathematics technique for quick multiplication. For any number N and a base B, if N is away from the base, the product can be calculated using the following formula:

$$N \times (B + \delta) = (N - \delta) \times (B - \delta) + \delta^2$$

Where δ is the deviation from the base.

Example:

Let's multiply 27 by 18 using the Nikhilam Sutra with the base 20:

$$27 \times (20 + 7) = (27 - 7) \times (20 - 7) + 7^2$$
$$= 20 \times 13 + 49$$
$$= 260 + 49$$
$$= 309$$

Therefore, $27 \times 18 = 309$.

2.4 Division

Division is the arithmetic operation that represents the process of distributing a quantity into equal parts or groups. It is denoted by the symbol $\nabla\cdot$.

2.4.1 Basic Division Formula

The basic division formula is represented as follows:

$$a\nabla \cdot b = c \tag{2.1}$$

where a is the dividend, b is the divisor, and c is the quotient.

2.4.2 Example: Long Division

Long division is a method commonly used to perform division of large numbers. Here's an example:

$$\frac{235}{7\,|\,1645} \tag{2.2}$$

In this example, 7 is the divisor, 1645 is the dividend, and the result is obtained step by step.

2.4.3 Fractional Division

Division can also result in fractional or decimal values. For example:

$$\frac{3}{4}\nabla \cdot \frac{1}{2} = \frac{3}{4} \times \frac{2}{1} = \frac{6}{4} = 1.5 \tag{2.3}$$

2.4.4   Division Properties

There are important properties associated with division, such as the division by zero is undefined.

2.4.5   Word Problem Example

Consider the following word problem:

Mary has 24 candies, and she wants to distribute them equally among her 6 friends. How many candies will each friend receive?

The division equation for this problem is $24\nabla \cdot 6 = 4$, so each friend will receive 4 candies.

Chapter 3

Advanced Techniques

3.1 Vedic Mathematics

Vedic Mathematics is a system of mathematical techniques that originated in ancient India. It offers a variety of methods for rapid mental calculations. One prominent technique is the process of multiplying two-digit numbers using the Vedic Sutras.

3.1.1 Two-Digit Multiplication: Vedic Sutras

The Vedic Sutra for two-digit multiplication is known as "Urdhva-Tiryagbhyam." The sutra states:

$$\text{Product} = \text{Left part} \times 100 + \text{Cross product} + \text{Right part}$$

The Left part is the product of the leftmost digits, the Right part is the product of the rightmost digits, and the Cross product is the product of the crosswise subtraction of each digit.

Example:

Let's multiply 23 by 34 using the Urdhva-Tiryagbhyam sutra:

$$\text{Left part} = 2 \times 3 = 6$$

$$\text{Right part} = 3 \times 4 = 12$$

$$\text{Cross product} = (2 + 4) \times (3 - 4) = -6$$

$$\text{Product} = 6 \times 100 + (-6) + 12 = 618$$

Therefore, $23 \times 34 = 618$.

3.1.2 Three-Digit Multiplication: Vedic Sutras

For three-digit multiplication, the "Nikhilam" sutra can be applied. The sutra states:

$$\text{Product} = (\text{First part} - \text{Deviation}) \times (\text{Base} - \text{Deviation}) + \text{Deviation}^2$$

The First part is the product of the leftmost digits, the Deviation is the difference between each digit and the chosen base (usually 10), and the Base is the chosen base for the calculation.

Example:

Let's multiply 327 by 214 using the Nikhilam sutra with a base of 100:

$$\text{First part} = 3 \times 2 = 6$$

$$\text{Deviation} = (3 - 10) + (2 - 10) + (7 - 10) = -3$$

$$\text{Product} = (6 - (-3)) \times (100 - (-3)) + (-3)^2 = 66 \times 103 + 9 = 6,798$$

Therefore, $327 \times 214 = 6,798$.

3.2 Trachtenberg System

The Trachtenberg System is a set of speed arithmetic techniques developed by Jakow Trachtenberg. It includes methods for multiplication, division, addition, and subtraction. In this section, we focus on the Trachtenberg multiplication method.

3.2.1 Trachtenberg Multiplication Method

The Trachtenberg Multiplication method is a systematic approach to multiply any two numbers. The method involves creating a matrix and filling it with intermediate values.

$$\begin{bmatrix} & A & B & C & D \\ \times & E & F & G & H \\ \hline & P & Q & R & S \\ & T & U & V & W \end{bmatrix}$$

The matrix is filled with values according to the following rules:

$$P = A \times E$$

$$Q = B \times E$$

$$R = C \times E$$

$$S = D \times E$$

$$T = A \times F$$

$$U = B \times F$$

$$V = C \times F$$

$$W = D \times F$$

The final result is obtained by combining the values from the matrix:

$$\text{Result} = 1000P + 100Q + 10R + S + 1000T + 100U + 10V + W$$

Example:

Let's multiply 327 by 214 using the Trachtenberg method:

$$
\begin{bmatrix}
 & 3 & 2 & 7 \\
\times & 2 & 1 & 4 \\
\end{bmatrix}
$$

Now, filling in the matrix:

$$P = 3 \times 2 = 6$$

$$Q = 2 \times 2 = 4$$

$$R = 7 \times 2 = 14$$

$$S = 7 \times 4 = 28$$

$$T = 3 \times 1 = 3$$

$$U = 2 \times 1 = 2$$

$$V = 7 \times 1 = 7$$

$$W = 7 \times 1 = 7$$

Now, combining the values:

$$\text{Result} = 1000 \times 6 + 100 \times 4 + 10 \times 14 + 28 + 1000 \times 3 + 100 \times 2 + 10 \times 7 + 7$$

$$= 6000 + 400 + 140 + 28 + 3000 + 200 + 70 + 7$$

$$= 9217$$

Therefore, $327 \times 214 = 9217$.

Chapter 4

Practical Exercises

4.1 Speed Drills

Speed drills are an integral part of mastering speed arithmetic. Regular practice enhances mental calculation abilities, allowing for quicker and more accurate computations. In this section, we will explore various speed drills for addition, subtraction, multiplication, and division.

4.1.1 Addition Speed Drill

For addition speed drills, a key approach is to practice adding sets of numbers in quick succession. Start with small numbers and gradually increase the difficulty. The goal is to improve mental agility and accuracy.

Example:

Perform the following addition speed drill:

$$256$$
$$+ \quad 789$$

Answer: 1045

4.1.2 Subtraction Speed Drill

Subtraction speed drills involve practicing subtracting numbers rapidly. Begin with simple problems and progress to more challenging ones. The aim is to enhance the ability to perform quick mental subtractions.

Example:

Perform the following subtraction speed drill:

$$
\begin{array}{r}
789 \\
-\ 456 \\
\hline
\end{array}
$$

Answer: 333

4.1.3 Multiplication Speed Drill

For multiplication speed drills, focus on multiplying two-digit or three-digit numbers quickly. The goal is to improve mental multiplication skills and efficiency.

Example:

Perform the following multiplication speed drill:

$$
\begin{array}{r}
23 \\
\times\ 34 \\
\hline
\end{array}
$$

Answer: 782

4.1.4 Division Speed Drill

Division speed drills involve practicing dividing numbers mentally. Start with basic division problems and progressively move to more complex ones.

Example:

Perform the following division speed drill:

$$\nabla \cdot \quad \frac{256}{8}$$

Answer: 32

4.1.5   Mixed Operation Speed Drill

Combine addition, subtraction, multiplication, and division problems for a comprehensive speed drill. This challenges the mind to switch between different operations rapidly.

Example:

Perform the following mixed operation speed drill:

$$(324 + 58) \times 2 - 137\nabla \cdot 7$$

Answer: 950

4.2   Real-world Applications

Speed arithmetic skills find practical applications in various real-world scenarios. From managing personal finances to making quick decisions in professional settings, the ability to perform mental calculations efficiently is invaluable.

4.2.1   Budgeting and Personal Finance

In everyday life, individuals often need to make quick financial decisions. The ability to mentally calculate expenses, plan budgets, and estimate savings is crucial. For example, consider the following scenario:

Example:

You are at a grocery store with a list of items and their prices. By mentally estimating the total cost, you can make informed decisions about your spending without the need for a calculator.

4.2.2 Work-related Calculations

In professional settings, speed arithmetic is beneficial for making quick decisions and solving problems on the spot. Whether it's estimating project costs, calculating percentages, or analyzing data, mental calculations save time.

Example:

During a business meeting, you need to quickly calculate the percentage increase in sales for a particular product. Speed arithmetic allows you to provide an answer promptly, demonstrating efficiency and competence.

4.2.3 Time Management

Efficient time management is essential in various aspects of life. Speed arithmetic helps individuals allocate time effectively, whether it's planning daily schedules, estimating travel times, or organizing tasks.

Example:

You have a series of tasks to complete throughout the day. By quickly estimating the time required for each task, you can create a realistic schedule, ensuring that you allocate sufficient time for each activity.

4.2.4 Shopping and Discounts

When shopping, the ability to mentally calculate discounts, compare prices, and determine the best deals is advantageous. Speed arithmetic enables individuals to make informed purchasing decisions.

Example:

You come across a discount offer with a percentage off the original price. Mentally calculating the discounted price allows you to evaluate whether the offer is a good deal.

4.2.5 Travel Planning

Whether planning a road trip or booking flights, speed arithmetic aids in estimating travel times, distances, and costs. It facilitates quick decision-making during travel arrangements.

Example:

You are planning a road trip and need to estimate the total travel time, considering distances between destinations. Speed arithmetic allows you to make informed decisions about the itinerary.

4.2.6 Health and Fitness

In health and fitness, individuals may need to calculate caloric intake, track exercise progress, or make quick nutritional decisions. Speed arithmetic supports these calculations.

Example:

You are following a fitness plan that requires calculating the total number of calories consumed in a day. Speed arithmetic helps you monitor your nutritional goals efficiently.

4.2.7 Problem-solving in Everyday Situations

Speed arithmetic fosters problem-solving skills, allowing individuals to tackle various challenges that arise in daily life. It encourages a quick and effective approach to finding solutions.

Example:

You encounter a situation where you need to distribute expenses among a group of friends. Speed arithmetic enables you to quickly determine each person's share.

Chapter 5

Speed Arithmetic Formulas

5.1 General Formulas

This section presents general formulas for various arithmetic operations that can be applied for quick mental calculations. These formulas aim to simplify the process and enhance computational speed.

5.1.1 Two-Digit Multiplication

For two-digit numbers AB and CD, the product can be calculated using the formula:

$$AB \times CD = (A \times C) \times 100 + (A \times D + B \times C) \times 10 + B \times D$$

Example:

Let's multiply 23 by 34 using the formula:

$$23 \times 34 = (2 \times 3) \times 100 + (2 \times 4 + 3 \times 3) \times 10 + 3 \times 4$$
$$= 782$$

5.1.2 Complementary Subtraction

Complementary subtraction is a technique to transform subtraction into addition. For numbers A and B where $A > B$, the subtraction $A - B$ can be calculated using the formula:

$$A - B = A + (10 - B)$$

Example:

Let's subtract 7 from 14 using complementary subtraction:

$$14 - 7 = 14 + (10 - 7)$$
$$= 17$$

5.1.3   Nikhilam Sutra for Multiplication

The Nikhilam Sutra is a Vedic Mathematics technique for multiplication. For any number N and a base B, if N is away from the base, the product can be calculated using the formula:

$$N \times (B + \delta) = (N - \delta) \times (B - \delta) + \delta^2$$

Where δ is the deviation from the base.

Example:

Let's multiply 27 by 18 using the Nikhilam Sutra with the base 20:

$$27 \times (20 + 7) = (27 - 7) \times (20 - 7) + 7^2$$
$$= 309$$

5.2   Application Examples

This section presents practical application examples of speed arithmetic formulas. Each example includes working examples and numerical calculations.

5.2.1   Example 1: Rapid Budget Estimation

Consider a scenario where you are grocery shopping and want to estimate the total cost quickly. The Two-Digit Multiplication formula can be applied.

Scenario:

You have four items with prices $12, 18, 25,$ and 30. Calculate the estimated total cost.

$$12 \times 4 + 18 \times 3 + 25 \times 2 + 30 \times 1 = 48 + 54 + 50 + 30$$
$$= 182$$

The estimated total cost is 182.

5.2.2 Example 2: Complementary Subtraction in Shopping

In a shopping scenario, you want to quickly find the price difference between two items. Complementary subtraction can be employed.

Scenario:

The original price of an item is $47,$ and it is currently on sale for 32. Calculate the price difference using complementary subtraction.

$$47 - 32 = 47 + (10 - 32)$$
$$= 47 + 18$$
$$= 65$$

The price difference is 65.

5.2.3 Example 3: Speedy Multiplication with Nikhilam Sutra

Suppose you need to calculate the total cost of multiple items with varying quantities. The Nikhilam Sutra for multiplication can expedite the process.

Scenario:

You are purchasing 5 items priced at 14 each. Estimate the total cost using the Nikhilam Sutra with a base of 10.

$$5 \times (10 + 4) = (5 - 4) \times (10 - 4) + 4^2$$
$$= 24$$

The estimated total cost is 24.

5.2.4　Example 4: Efficient Division in Cooking

In a cooking scenario, you want to adjust a recipe for a different number of servings. The complementary division method can be useful.

Scenario:

A recipe is designed for 4 servings, and you want to adjust it for 8 servings. Use complementary division to find the adjusted quantities.

$$\frac{4}{8} = \frac{4}{4 + (10 - 8)}$$
$$= \frac{4}{6}$$
$$= \frac{2}{3}$$

The adjusted quantity is $\frac{2}{3}$ of the original.

Chapter 6

Memory Techniques

6.1 Mnemonics

Mnemonics are memory aids that help individuals remember information more effectively. In speed arithmetic, mnemonics can be used to recall formulas, sequences, and other mathematical concepts quickly. This section explores various mnemonic techniques and their applications.

6.1.1 Acronyms for Formulas

Creating acronyms is a useful mnemonic technique for remembering formulas. By forming a memorable word or phrase using the first letters of the formula components, one can recall the formula more easily.

Example:

Consider the formula for finding the area of a rectangle: $A = l \times w$. An acronym for this formula could be "ALW," where A stands for Area, L for Length, and W for Width. This makes it easier to remember the formula when needed.

6.1.2 Rhymes and Jingles

Rhymes and jingles are memorable patterns of sound that can aid in recalling information. Creating a catchy rhyme or jingle for a formula or concept can make it stick in memory.

Example:

To remember the order of operations in mathematics (PEMDAS: Parentheses, Exponents, Multiplication and Division, Addition and Subtraction), one might use a jingle: "Please Excuse My Dear Aunt Sally." Each word corresponds to a step in the order of operations.

6.1.3 Visual Associations

Creating visual associations between numbers or mathematical concepts and familiar images can enhance memory retention. This technique involves mentally connecting an image with the information to be remembered.

Example:

To remember the multiplication table for 9, visualize fingers numbered 1 to 10. When multiplying by 9, fold down the corresponding finger to the number being multiplied. The number of fingers before the folded one represents the tens digit, and the number of fingers after represents the ones digit.

6.1.4 Chunking and Grouping

Chunking involves breaking down a large amount of information into smaller, manageable chunks. Grouping similar items together aids in organizing and recalling the information more efficiently.

Example:

When memorizing a sequence of digits, such as a phone number, group the digits into smaller chunks. For example, instead of remembering "1234567890," remember it as "123-456-7890."

6.1.5 Storytelling and Narratives

Creating stories or narratives around mathematical concepts can make them more memorable. By associating information with a story, individuals can recall the details more effectively.

Example:

To remember the quadratic formula $(ax^2 + bx + c = 0)$, create a story where the coefficients a, b, and c have specific roles or characters. The story provides context and aids in remembering the formula.

6.1.6 Musical Mnemonics

Associating mathematical concepts with musical patterns or tunes can enhance memory retention. Creating a musical mnemonic for a formula or sequence can make it more engaging and memorable.

Example:

Compose a short melody for the Pythagorean theorem $(a^2 + b^2 = c^2)$. Singing or mentally replaying the melody can help recall the formula during calculations.

6.1.7 Spatial Memory Techniques

Utilizing spatial memory involves associating information with specific locations or positions. Creating mental maps or spatial arrangements aids in recalling details.

Example:

To remember the trigonometric ratios (SOH-CAH-TOA), visualize a right-angled triangle with the sides labeled appropriately. The spatial arrangement reinforces the relationships between the sides and angles.

6.1.8 Keyword Mnemonics

Associating keywords or trigger words with mathematical concepts can serve as effective memory cues. These keywords act as shortcuts to remember specific details.

Example:

To remember the sine, cosine, and tangent in trigonometry, use the mnemonic "Some Old Hens Can Always Hop To Our Amazement." The initial letters correspond to the trigonometric ratios.

6.1.9 Repetition and Practice

Repetition is a fundamental mnemonic technique. Regular practice reinforces memory, making it easier to recall information when needed.

Example:

Repeatedly practicing mental calculations using specific formulas or techniques solidifies the memory. The more frequently the information is revisited, the stronger the memory becomes.

6.1.10 Summary

Mnemonics play a crucial role in enhancing memory and recall in speed arithmetic. These techniques provide creative and engaging ways to remember formulas, sequences, and mathematical concepts. Whether through acronyms, rhymes, visual associations, or other methods, mnemonics contribute to the development of mental agility in arithmetic.

Chapter 7

Speed Arithmetic Challenges

7.1 Competitions

Speed arithmetic competitions are a thrilling way to showcase and enhance one's mental calculation skills. These challenges often involve solving complex problems within strict time constraints. In this section, we will explore various types of speed arithmetic competitions, along with strategies and formulas that can be applied.

7.1.1 Single-digit Addition Sprint

In this competition, participants are given a series of single-digit addition problems to solve within a limited time. The key is to quickly add the digits without writing down intermediate steps.

Strategy:

Employ mental addition techniques, such as grouping numbers to make tens or utilizing complementary addition, to enhance speed. For example, when adding

$7 + 8$, think of it as $7 + 3 + 5$, making a group of 10.

Example:

Solve the following problems in 30 seconds:

$$
\begin{array}{r}
5 + 9 + 2 \\
+\ 4 + 7 + 6 \\
\hline
\end{array}
$$

7.1.2   Multiplication Madness

Participants compete in solving multiplication problems involving two or three-digit numbers. The challenge is to compute the products rapidly without the aid of calculators or written calculations.

Strategy:

Utilize mental multiplication techniques such as the Nikhilam Sutra or break down the multiplication into simpler steps. For instance, when multiplying 23 by 15, break it into $(20 + 3) \times 15$.

Example:

Solve the following problems in 1 minute:

$$
\begin{array}{r}
23 \ \times \ 15 \\
\times \ 17 \ \times \ 12 \\
\hline
\end{array}
$$

7.1.3   Rapid Mental Square Roots

Competitors are challenged to calculate square roots mentally for perfect squares. The goal is to swiftly determine the square root without resorting to written methods.

Strategy:

Use estimation and approximation to quickly identify the closest perfect square. Apply the knowledge of squares and square roots for rapid mental calculations.

Example:

Solve the following problems in 45 seconds:

$$\sqrt{64}$$
$$\overline{\sqrt{121}}$$

7.1.4   Decimal Division Dash

In this competition, participants tackle division problems involving decimals. The challenge is to perform precise mental divisions within a stipulated time.

Strategy:

Utilize complementary division and approximation techniques to simplify the division process. Break down complex divisions into smaller, more manageable steps.

Example:

Solve the following problems in 1 minute:

$$\frac{45.6}{2}$$
$$\overline{\frac{78.9}{3}}$$

7.1.5   Mixed Operation Marathon

Participants face a barrage of mixed arithmetic problems involving addition, subtraction, multiplication, and division. The challenge is to solve a series of problems within a tight timeframe.

Strategy:

Develop a systematic approach to switch between different operations rapidly. Prioritize problems based on their complexity and potential for quick mental solutions.

Example:

Solve the following problems in 2 minutes:

$$(45 + 18) \times 3 - \frac{72}{4}$$

7.1.6 Summary

Participating in speed arithmetic competitions not only tests one's mental agility but also enhances the ability to perform complex calculations swiftly. Developing effective strategies and practicing mental arithmetic regularly are key to success in these challenges. The formulas and techniques presented in this book serve as valuable tools for tackling the diverse range of problems encountered in speed arithmetic competitions.

Chapter 8

Conclusion

8.1 Summary

In this comprehensive handbook on speed arithmetic, we have delved into a myriad of formulas, techniques, and strategies to enhance mental calculation skills. The journey through the various chapters has equipped readers with tools to perform rapid and accurate arithmetic in diverse scenarios.

8.1.1 Basic Arithmetic Foundations

The book began by establishing a solid foundation in basic arithmetic operations such as addition, subtraction, multiplication, and division. Through techniques like complementary addition and subtraction, as well as the Vedic Mathematics approach, readers gained insights into streamlining fundamental calculations.

8.1.2 Advanced Techniques for Multiplication and Division

Building on the basics, advanced techniques for multiplication and division were explored. The Nikhilam Sutra, spatial memory, and complementary division methods provided powerful tools for tackling more complex arithmetic problems with ease.

8.1.3   Specialized Topics: Vedic Mathematics and Trachtenberg System

Diving into specialized topics, we explored the Vedic Mathematics system and the Trachtenberg System. These ancient and modern methodologies respectively provided unique perspectives on arithmetic, offering alternative approaches to traditional calculation methods.

8.1.4   Memory Techniques and Mnemonics

Recognizing the role of memory in arithmetic prowess, we delved into mnemonics and memory techniques. Acronyms, rhymes, visual associations, and other mnemonic strategies were presented to aid in recalling formulas, sequences, and mathematical concepts.

8.1.5   Practical Applications and Real-world Scenarios

The handbook did not merely focus on theoretical concepts but also emphasized practical applications. Real-world scenarios, such as budget estimation, shopping calculations, and cooking adjustments, were used to demonstrate the applicability of speed arithmetic in everyday life.

8.1.6   Competitions and Challenges

To put acquired skills to the test, the book covered various speed arithmetic competitions. From single-digit addition sprints to mixed-operation marathons, readers were exposed to challenges that require quick mental calculations under strict time constraints.

8.1.7 Conclusion of the Journey

As we conclude this journey through the world of speed arithmetic, it is essential to reflect on the key takeaways. Mental agility, strategic thinking, and a systematic approach to problem-solving are the cornerstones of mastering speed arithmetic.

8.1.8 Continuous Learning and Practice

Speed arithmetic is not a destination but a continuous journey of learning and refinement. Regular practice and exposure to diverse problem types contribute to the development of mental calculation skills.

8.1.9 Empowerment Through Mathematics

Speed arithmetic empowers individuals by enabling them to perform complex calculations mentally and efficiently. This skill is not only valuable in academic and professional settings but also enhances problem-solving abilities in various aspects of life.

8.1.10 Encouragement to Explore Further

As readers conclude this handbook, the encouragement is extended to explore further. Delve into additional mathematical concepts, discover personal shortcuts, and continue refining mental arithmetic skills.

8.1.11 Gratitude and Acknowledgments

Finally, heartfelt gratitude is extended to all readers who embarked on this journey. The pursuit of speed arithmetic is an intellectual adventure, and the author sincerely hopes that the knowledge shared in this handbook contributes to the enrichment of mathematical abilities.

8.1.12 May Your Calculations Be Swift and Accurate!

In the spirit of speed arithmetic, may your calculations be swift and accurate. With the acquired formulas, techniques, and strategies, embrace the world of mental mathematics with confidence and precision.

8.2 Looking Forward

As we conclude this comprehensive exploration of speed arithmetic, it is crucial to consider the road ahead. The journey through various formulas, techniques, and applications has provided a solid foundation for mastering mental calculation skills. Looking forward, there are several avenues to explore and further refine one's ability to perform rapid and accurate arithmetic.

8.2.1 Integration of Advanced Techniques

Building on the advanced techniques presented in this handbook, the path forward involves integrating these methods seamlessly into daily mathematical tasks. Practice incorporating the Nikhilam Sutra, Vedic Mathematics principles, and Trachtenberg System into diverse problem-solving scenarios to enhance proficiency.

8.2.2 Exploration of Additional Speed Arithmetic Systems

Speed arithmetic is a vast field with multiple systems and approaches. The future holds the opportunity to explore additional systems, each offering its unique insights and shortcuts. Consider delving into less mainstream methods to uncover hidden gems that resonate with personal learning preferences.

8.2.3 Application in Specialized Fields

While the handbook covered practical applications in everyday scenarios, looking forward involves exploring the application of speed arithmetic in specialized fields. Investigate how these mental calculation skills can be leveraged in professional settings such as finance, engineering, computer science, and more.

8.2.4   Incorporation of Technology

As technology continues to advance, the integration of digital tools and applications for speed arithmetic is an exciting prospect. Explore how computational tools can complement and enhance mental arithmetic skills, fostering a symbiotic relationship between human calculation and technological assistance.

8.2.5   Teaching and Sharing Knowledge

One of the most rewarding paths forward is to share the acquired knowledge with others. Consider becoming an advocate for speed arithmetic education, whether through tutoring, workshops, or online platforms. Empowering others with these valuable skills contributes to the broader dissemination of mathematical proficiency.

8.2.6   Continuous Practice and Challenges

The journey of speed arithmetic is an ongoing process of improvement. Looking forward involves setting new challenges, both in terms of complexity and speed. Regular practice, exposure to diverse problem types, and participation in speed arithmetic challenges contribute to continuous refinement and growth.

8.2.7   Integration with Lifelong Learning

Speed arithmetic is just one facet of the broader landscape of lifelong learning. As learners embark on various educational pursuits, the integration of mental arithmetic skills enhances problem-solving capabilities across different domains. Cultivate a mindset of continuous learning and adaptation.

8.2.8 Exploration of Personal Shortcuts

Individuals often discover personal shortcuts and strategies that resonate with their unique thought processes. Looking forward invites exploration into these personal discoveries, fostering a personalized approach to speed arithmetic that aligns with individual cognitive strengths.

8.2.9 Community Engagement and Collaboration

Engage with the speed arithmetic community and collaborate with like-minded enthusiasts. Share insights, exchange tips, and collectively contribute to the evolution of speed arithmetic methodologies. The synergy of collaborative efforts can lead to the discovery of novel techniques and approaches.

8.2.10 Gratitude for the Journey

As we look forward to the future of speed arithmetic, it is essential to express gratitude for the journey thus far. The pursuit of mental calculation skills is an enriching endeavor that extends beyond the pages of this handbook. May the knowledge gained pave the way for a future filled with mathematical curiosity and accomplishment.